Doris Breiner

Handbook: The Duty for Sponsor Oversight in Clinical Trials
Supplement to the e-book Edition 1 and 2

Doris Breiner

HANDBOOK: THE DUTY FOR SPONSOR OVERSIGHT IN CLINICAL TRIALS

SUPPLEMENT TO THE E-BOOK EDITION 1 AND 2

A PRACTICAL GUIDE: "RISK-BASED OPERATIONAL DATA REVIEW"

Impressum

Bibliographic information from the German National Library: The German National Library lists this publication in the German National Bibliography; detailed bibliographic data are available on the Internet at http://dnb.dnb.de.

The automated analysis of the work in order to obtain information, in particular about patterns, trends and correlations pursuant to Section 44b of the German Copyright Act ("Text and Data Mining"), is prohibited.

© 2024 Doris Breiner

Publisher: BoD · Books on Demand GmbH, In de Tarpen 42, 22848 Norderstedt, bod@bod.de
Printing: Libri Plureos GmbH, Friedensallee 273, 22763 Hamburg

ISBN: 978-3-7693-1922-4

Bibliografische Information der Deutschen Nationalbibliothek: Die Deutsche Nationalbibliothek verzeichnet diese Publikation in der Deutschen Nationalbibliografie; detaillierte bibliografische Daten sind im Internet über http://dnb.dnb.de abrufbar.

Die automatisierte Analyse des Werkes, um daraus Informationen insbesondere über Muster, Trends und Korrelationen gemäß §44b UrhG („Text und Data Mining") zu gewinnen, ist untersagt.

© 2024 Doris Breiner

Verlag: BoD · Books on Demand GmbH, In de Tarpen 42, 22848 Norderstedt, bod@bod.de
Druck: Libri Plureos GmbH, Friedensallee 273, 22763 Hamburg

ISBN: 978-3-7693-1922-4

Preamble

The next edition of the practical guide with regard to "The Duty for Sponsor Oversight in Clinical Research" was prepared to present in more detail the "Risk Based Operational Data Review". As outlined in the previous edition of the practical guide, the definition of a threshold for missing data and or values was not explicit outlined, in the applicable regulatory binding documents, for example the Clinical trials - Regulation EU No 536/2014. This is also applicable for the upper limit or level of acceptance of, for example, protocol deviations, missing values and or values out of rang or missing safety visits. Nevertheless, the underlying regulations as well as reflexion papers and other articles provide the overall guidance, and required information for the determination of relevant protocol deviations, serious breaches and other deviations or issues. These should normally be aligned to the defined outcome parameters of a clinical trial. The same applies for the patient safety, data integrity and data protection Therefore, the supplement edition was prepared for proposing a separate risk-based operational review.

Considering, for example, the clinical trial phase, the frequency of the protocol deviations as well as the information the protocol deviation could deliver, seems a reasonable approach. Furthermore, the revised EU-Commission AI Act Regulation (EU) 2024/1689, entered into force on 01 August 2024, was considered regarding the qualification procedures for service providers. Especially, if pre-programmed review tools would be established, the quality assurance procedure should include an assessment of, for example, the validated data exports. This process should include a risk-based assessment of the AI-assisted application provider before any application will be used. It applies also for the manufacturing, distribution of an investigational medicinal product. For covering this aspect in advance and due to the international cooperations, an example for the determination of a pre-study risk score assignment will be shown in Table 1. The edition outlines in Table 2 the chapter 4 of the ICH-GCP and assigned investigator is responsibilities. Nonetheless, the sponsor remains as the overall responsible party for the clinical trial conduct and proper analyses according to the underlying regulatory requirements. These can be found in chapter 5 of the ICH-GCP, and are summarised in the Table 3. *This edition could be categorised as a supplement to the 2nd edition.* Nevertheless, a separate guide was prepared to demonstrate an approach for the clinical trial site risk-based data and protocol deviation review. The aim includes enhancing with the assumed procedure the quality of the data-set as required, for example, the AMNOG procedure in Germany and or accelerated approval.

Also, any changes related to the topic which induced any correction, with regard to the previous editions, are summarised briefly below. Finally, the current revision of the E6 ICH-GCP R2 (R3) will be noted. The anticipated finalisation of the revised guidance was determined for 2025.

Berlin, 12 December 2024 Doris Breiner, MSc

Outcome of further discussions over the time, with regard to the content of the 2nd edition, summarised briefly.

General: The revised EU-Commission AI Act Regulation (EU) 2024/1689, made available on 01 August 2024, *(Reference 29),* was considered regarding the use of artificial intelligence information technology in clinical trials. Especially, it concerns partially automated review and or documentation processes. The regulation outlines the high-risk categories for protecting and ensuring the safeguard of the user, environment and manufacturer. Furthermore, a revised guidance was provided by the Food and Drug Administration in April 2023, reinforcing the importance of Sponsor Oversight and Quality Risk Management in All Clinical Trials *(Reference 30)*.

In addition to that, the assumed changes of the E6 ICH-GCP R2 (R3) will be briefly remarked. According to the most recent information, the final version could be expected in 2025 *(Reference 32)*.
The terminology for the "Fast Track Approval" was replaced by "Accelerated Approval" at least by as example the Food and Drug Administration (USA).

Chapter 5:
The language for the example with regard to the legend for scoring criteria was slightly adjusted.
A few small spelling corrections in several chapters, for example, the use of capital letters and hyphen, but also the alignment of extra signs formatting, according to the requirements of the provider.

.

TABLE OF CONTENT

1.0. Introduction

The practical guide specifies in the following edition in more detail the "Risk Based Operational Data Review" as part of the clinical data review and sponsor oversight activities. *First, a brief summary related to the previous e-book edition 1 and 2 of the practical guide can be found.* The access for the marketing authorization of new pharmaceutical products requires specified data, including the proper clinical data documentation. This also applies for further established treatment options of authorized medicinal products. As presented before, the applicable regulatory requirements are outlined in the Clinical trials - Regulation EU No 536/2014 *(Regulation (EU) No 536/2014))*[1], the German Drug Law [2], the Clinical trials Directive 2001/20/EC [3], and the German GCP-V Verordnung [4]. The Regulation (EU) No 536/2014 was established end of January 2022. The requirements for the manufacturing of an investigational medicinal product (IMP) according to the Good Manufacturing Practice (GMP) [5] as well as the requirements for the AMNOG procedure, are very specific [6], [7].

Therefore, the practical guide considers risk-based quality control activities and measures addressing the GMP and Good Clinical Practice (GCP) [8]. The consideration of these presented quality aspects could contribute to providing a comprehensive, consistent data-set, for ensuring the applicable requirements for the AMNOG assessment as required by the G-BA are covered [6] (Gemeinsamer Bundesausschuss). As defined in the "International Council for Harmonisation of Technical Requirements for Pharmaceuticals for Human Use - E6 (R2) Good clinical practice" (ICH-GCP) the investigator is responsible for the fulfilment and adherence to the requirements as outlined in the chapter 4 [8]. Nevertheless, the overall responsibility for the clinical trial and data remains with the sponsor and is therefore part of the sponsor oversight, as outlined in the ICH-GCP chapter 5.2 [8]. The supplemental edition of the practical guide specifically outlines a proposal for the conduct of the "Risk Based Operational Data Review" and includes also the clinical trial site and protocol deviation assessment. This review corresponds to the multifactorial component 1 as previously described in the 2nd edition of the practical guide.

For consistency reasons, the term "Risk Based Operational Data Review"[1] was selected, but does not reflect a standardised abbreviation. In the follow, the altered term *separate operational review* will be used as well. For the development of a scale with pre-defined selection

---

[1] The term "Risk Based Operational Data Review" was selected for consistency and covers the "Risk Based Clinical Trial Site and Protocol Deviation Review". This term does not reflect a standardised abbreviation. The altered term *separate operational review* will be used as well.

criteria (for example criteria related to the definition of a score) the 5 Point Likert was considered. This scale analysis is frequently used and outlines answers assumed supporting a stable analyses [9]. Additionally, the investigation provided by K. Getz, *"Quantifying Protocol Deviation Experience by Clinical Phase"* [10] was considered. The article points out that limitations of the protocol deviations assessment are commonly seen. It was noted minimising the amount of missing relevant data should be paid attention for ensuring the completeness of the clinical data-set. Therefore, the article *"Development of a standardised set of metrics for monitoring site performance in multicentre randomised trials: a Delphi study"* [11] was seen as supportive. Figure 2, shows a practical example for a risk-based quality site performance metrics. It could allow the appropriate preventive analyses for addressing the issue. Especially, defining a threshold and or level of acceptance for missing values and or assessments was not concretely named. The same information could be found, for example, in the scientific guideline *"Missing data in confirmatory clinical trials"* - Scientific guideline published by the European Medicines Agency (EMA) [12].

Furthermore, the recommendations published by the *"Transcelerate Group"*, [13], [14], [15] outline similar information regarding the definition of a threshold. Also, the observations regarding the management of protocol deviations, outlined in the summary paper *"MHRA Identifies Common Flaw in Managing Protocol Deviations"* [16], indicates the requirement for risk-based approaches. The previously referenced publication, prepared by Coens et al., *"The Setting International Standards in Analysing Patient-Reported Outcomes and Quality of Life Endpoints"* (SISAQOL) Consortium, provides a further proposal for the handling of missing values and data. With this approach, a proper analyses of the quality-of-life (QoL) endpoints and general health improvement could be ensured [17]. In particular, the method described in the article SISAQOL was seen as supportive for stabilising the proposed rationale presented in the practical guide. This is because the data will be taken into account for the AMNOG assessment but also for the ongoing risk-benefit-assessment.

In addition to that, the distribution of the suggested weighted value, for the *separate operational review*, as part of the multifactorial component score 1, was considered [18], [19].

As presented in the 2nd edition of the practical guide, the proposed selection criteria were determined to ensure the data items of relevance for the primary, secondary endpoints are covered. This also remains applicable for the safety data as well as the overall corresponding transparent documentation. For the ICH-GCP item 4.9, Records and Reports, the selection criteria for the review of the investigator site file and the documentation were based on the current available regulatory guidelines and recommendations [1], [2], [3], [4]. Furthermore, the recommendations published by the Medicines and Healthcare products Regulatory Agency (MHRA) GxP data integrity guide are providing relevant information [20]. The abbreviation GxP is referencing the various good practice regulations and guidelines, for example, the good laboratory practice, the good documentation practice (GDP) and good laboratory practice (GLP). Protocol deviations could occur in all GxP related activities [21], [22] [23]. The discussions about this topic are always subject within professional international meetings. As pointed out before, it depends on the policies and standardised procedures established within the quality system of a sponsor organisation [1], [2], [3], [4].

Also for this separate operational review, a partially automated programming, for example, by using artificial intelligence (AI) applications could be considered [24], [25]. Nonetheless, the regulatory requirements need to be ensured. Also, the recommendations regarding the risk-based quality management as provided by the EMA [26], and ICH [27], [28] were considered for the use of partial technical assisted review activities. The revised European Union-Commission AI Act Regulation (EU) 2024/1689 *(Regulation (EU) 2024/1689),* published on 01 August 2024, [29] should therefore be taken into account. In particular, the outlined categorisation into risk groups should, among other things, guarantee the protection of fundamental rights, especially for high-risk-based AI systems. Furthermore, the revised guidance prepared by the Food and Drug Administration (FDA) in April 2023 [30] points out the reinforcing and importance of sponsor oversight and quality risk management in all clinical trials. In addition to that, any possible advantage of an ongoing analyses of this assessment by the data science discipline was discussed and considered as well [31].
The proposed approach was categorised as an appropriate, supportive measure for the guarantee of the adherence with the regulatory requirements. It would also enable identifying early on a gap in the infrastructure at the clinical trial site but as well in the quality system of a sponsor organisation [26], [27], [28].

Finally, the assumed update of the E6 ICH-GCP R2 (R3) document should be noted. The described changes include information regarding non-traditional clinical trial design, as well as decentralised trials. Also, the proportionality for the implementation of an efficient risk-based approach was addressed. The draft version was endorsed on 19 May 2023 and released for public consultation (step 2b) but not yet finalised [32].

2.0. Underlying Rationale and Assumptions to the Approach

In principle, the rationale for the oversight activities as provided in the previous editions remains the same. For the "Risk Based Operational Data Review" several different aspects should be considered. The retrospective reconstruction of a clinical trial, the impact on the outcome data, the data integrity, data protection and or patient safety remains the point of attention during regulatory inspections and or audits [1], [27], [28], [30]. Ensuring transparency in the documentation of protocol deviations, serious breaches as well as other deviations, should be ensured for the adherence to the underlying regulatory requirements [21], [22], [23].

At least approximately 75 % – 80 % of all different tasks related to the clinical trial site activities would be recommended for the conduct of *the separate operational review*. Each clinical trial site, actively recruiting patients should at least be reviewed at two time-points, whereas for the assigned monitor a frequency of at least once a year could apply. Nonetheless, it depends on the strategy and processes of the company and especially the established risk-based quality control policy and or standardised procedures. The definition of selection criteria ensuring the distribution of the data items in a risk-based manner would also be recommended for the multifactorial component 1. The proposed activities were, as outlined in the chapter 1, aligned to the regulatory requirements, the reflection papers provided as additional guidance by regulatory agencies [12], [26]. In addition to that, professional discussions and recommendations, published by other authors, were taken into account as well [10], [11], [17].

A further literature research, considering the clinical trial site data review, was performed at two different time-points. The search criteria included the articles and or publications made available between 2017 and 2022. The priority was set on meta-analyses and systematic reviews. The availability of meta-analyses publications could not be confirmed. In overall, these professional discussions provide similar conclusions. Furthermore, the recommendations provided by the *"Transcelerate Group"*, as referenced in the chapter introduction, were considered. These include the procedures proposed for the risk-based handling of the protocol deviations [13], [14], [15].

The aim of the separate supplement to the practical guide should be classified as supportive for the overall clinical data review. Furthermore, considering the best practice approach for assessing and calculating the GCP conform clinical trial site compliance induced the separate publication review. This is also required for the reporting of the results to the regulatory authorities [1]. The schematic overview of the clinical data review flow was adjusted, presenting the additional information with regard to the *separate operational review*. This overview can be found on the cover page. It concerns the cumulated multifactorial component score 1 (items 1; 1.1; 1.2) including the assigned weighted value as shown in the Table 5 [18], [19]. Depending on the internal responsibility matrix the conduct of the task, including the reporting of the review outcome, could be specified, for example, in a flowchart as an attachment to the quality plan. Especially, the different used quality review tools should be documented and outlined transparent and traceable in the quality plan and or superior quality management guidance document or manual [1].

In addition to the presented overall procedure the risk-based review and analyses of the protocol deviations, serious breaches would depend, for example, on the business outsourcing strategy of the company. For ensuring the sponsor oversight activities, an overall tracking, monitoring and analyses of the different deviations should also be established at the responsible sponsor organisation. Further analyses contributing to the ongoing assessment of the company risk-based quality system could be established, as required, and should be regular revised [11], [26], [28]. For the categorisation of the deviations different terms can be found, such as informative, non-informative, relevant or not relevant. In the common practice, the language major or minor for the categorisation of protocol deviations will be used. It should distinguish if an impact on the patient safety, data integrity and or data protection rules would be expected or not. Furthermore, the impact, possibly limiting the quality of the primary and secondary outcome measures, should be assessed. The review and categorisation would also include protocol deviations, serious breaches and other issues related to the use and management of an investigational medicinal product [1], [5], [8]. The guidance and recommendation documents offer the possibility of adjusting the term used for the categorisation of the protocol deviations [16], [20], [26]. The proposed following three categories, informative, potential informative and non-informative should cover these aspects. Other deviations would be part of the non-informative and or non-relevant deviations. The documentation of the observations requires the traceability for the assignment to the pre-defined criteria. The corresponding example for the legend ensuring a standardised approach for the assignment to the different categories can be found in the chapter 5. As remarked above, for the proposed criteria the principles of the 5 Point Likert were considered [9].

Protocol deviations and other deviations could occur in any stage of a clinical trial and or development program, including the information technology systems and other clinical trial material used. The risk-based approach demands the establishment for the proactive identification of anticipated risks. This would also concern the protocol deviations and or deviations due to the clinical trial design, processes and or the infrastructure of cooperation partners involved in the conduct of the clinical trial. Therefore, the conduct of a clinical trial in countries outside the EU, for example, other countries located in the Asia Pacific regions, would still require considering the national laws. The different locally applicable laws and or guidance documents would trigger the content of the contractual agreements [1]. This could induce, for example, in the United States, England and other countries, the confirmation of the content by a separate group, due to the national health care conditions and or data security and protection rules. Therefore, any deviations expected, for example, due to the business model and or local guidance with a cooperation partner should be part of a pre-study risk score assessment [16], [21], [22], [23]. Addressing the assumed general risks, possibly induced by the national law and or business model of the potential partners, would therefore be recommended in the first discussions of a clinical trial project. The corresponding examples can be found in the Table 1 "Criteria for the Pre-Study Risk Score".

In general, the contractual agreement and or any supplemental document should ensure the specific items are addressed accordingly. This also concerns conditions related to the manufacturing process of an investigational medicinal product, and or active pharmaceutical ingredient (API). A superior guiding document, for example, in a service agreement should be established, if applicable. For additional information provided in under other an IMP handling manual, the applicable regulatory legally binding document should appropriately be referenced. Inconsistencies between the regulatory binding underlying requirements, contractual agreements, and other customised instructions could also induce the occurrence of the different protocol deviations. In case of a legal conflict, resulting due to issues caused by, for example, non-adherence to the procedures by cooperation partners, the content of the contractual agreement remains the legally binding document [1], [5], [8].

In addition to that, the assumed risks potentially caused by the use of the AI-assisted applications, for example, established in information technologies should be paid special attention. The announcement for the final text of the Regulation (EU) 2024/1689 was made available on 01 August 2024 [29]. It demands the categorisation into risk groups among other things, to guarantee the protection of fundamental rights, especially for high-risk-based AI systems. The national implementation of the regulation would be expected within 12 months after the final release. Whereas, the reporting of possibly safety related incidences should be ensured within 6 months.

For business partners using AI applications, written evidence for the compliance with the GCP, GMP, as well as GDP, would be required and should be part of the pre-assessment report. Especially, any impact on the outcome data and or data integrity including data protection rules due to the used technical information AI applications should be considered [20], [21], [22], [23].

For the clarification of reporting requirements regarding possible AI-assisted applications related incidences a general scientific advice, at least for an assumed development program, could be considered. The national established reporting procedures should be referenced in, for example, the clinical trial protocol [1], [29]. For that purpose, the press release dated 08 May 2024, and published by the independent data protection conference, Germany could provide more information (Federal Commissioner for Data Protection and Freedom of Information (BfDI), and the state data protection authorities).

As initially outlined, the practical guide was based on the master thesis (2019). The handbook, "The Duty for Sponsor Oversight", draws attention to the practical approach of this responsibility. In comparison to the assumed overall clinical data review of at least approximately 75 % – 80 %, the *separate operational review* includes a review of all protocol deviations and or serious breaches occurred during the course of a clinical trial. As pointed out by the *"Transcelerate Group"* [13] the guidance "ICH E3 Structure and content of clinical study reports" demands the reporting of the performed data quality assurance (item 9.6) activities in the clinical study report (CSR).

Therefore, the regular reporting of the review would be recommended to be aligned to the underlying binding regulatory requirements. It should also present the rationale for the review schedule of the assigned data items, protocol deviation review, as well as the interpretation of the outcome data [33].

The aim of the supplemental edition remains, supporting the overall clinical data review by a separate risk-based quality assessment for the operational conduct of a clinical trial. The corresponding information can be found in the Table 5 (items 1; 1.1; 1.2) including the assigned weighted value for the multifactorial component, score 1. The weighted value was further divided and assigned to the applicable quality tools. Item 1.3 purely concerns the medical data review. The value of these data, as shown in chapter 3.2, provides the calculation and assignment of the score 1 [18], [19].

Selection Criteria for the "Risk-Based Operational Data Review" [1], [8], [26], [18], [19].

1. Clinical trial sites (at least all sites actively recruiting at any time during the clinical phase of the clinical trial).
2. Number of patients recruited and treated at the clinical trial site.
3. Protocol deviations, serious breaches and other deviations and or issues.
4. Monitoring report review, serious breaches and other deviations and or issues.
5. Co-monitoring visit at the clinical trial site and or remotely.
6. Pre-study cross-check between the guidance documents, contractual agreements and, for example, AI-assisted applications (refer to Table 1).

Additional note: AI-assisted functionalities assumed for the management and archiving of the clinical data and related documentation the suitability of the application should be assessed during the feasibility for the clinical trial.

Definition: Outcome of the site performance assessment (Score: 1 = good, 2 = medium, 3 = poor, based on the defined quality criteria).

The proposed five different aspects were kept, as presented in the previous 2nd edition, and classified as relevant information for the data integrity. The supportive description for the items 1-5 was adjusted accordingly. The item 6 was newly added for covering the relevant pre-study risk score assignment aspects. It also concerns the pre study identification of inconsistencies between the different clinical trial documents. The requirement for the sponsor oversight includes the release and authorization of clinical trial specific documents by the assigned sponsor representative. Furthermore, the pre-assessment of AI-assisted information technology regarding the adherence with the regulatory requirements should be considered [1], [5], [8], [29]. As pointed out above, the tracking tool for the documentation of the clinical trial should start with the first discussions with any experts and or assumed cooperation partners. Furthermore, it could be of interest, the tracking would distinguish between the source, the different clinical trial phases the different deviations, the serious breaches and other issues were identified. As outlined in the 2nd edition, the defined data items could be categorised, for example, as risk-based stratification factors [8], [16].

The retrieved score for the pre-identified overall risks should indicate the initial risk score, based on the design of the clinical trial and assumed procedures. The weighted score value, assigned to the protocol deviations, serious breaches and other deviations and or issues, was proposed for being further divided. The proposed percentages can be found in the Table 5. For ensuring a standardised assessment procedure, the source of a deviation was identified and or documented was considered as well. It was assumed the assignment percentage value could reflect a stable average outcome score [18], [19].

In addition to that, the distribution of the selected data items over the different review time-points could possibly be adjusted depending on the outcome of the review. The aim includes, ensuring an ongoing risk-based procedure for enhancing the quality of the research projects [10], [12]. Also, the identification of trends in the overall data-set should be more reliable with the proposed procedure. Below, examples for each selection criteria, also named as quality tool, can be found. These could be adjusted, as required, according to the company internal requirements and or internal standard quality procedures. As pointed out before, it remains with each sponsor establishing a rationale for a threshold and or upper level of acceptance for deviations [1], [26], [27], [28].

Independent of this, the impact of the deviation reporting on the calculated budget for the clinical trial proposal and clinical trial site would be of relevance. The judgement of the protocol deviations related to the recruitment should be reviewed by taking into account the initial assumed treatment exposed patients per clinical trial site, and or country.

Hence, as described above, the outcome summary of the complete clinical and quality review should according to the requirements be part of the CSR. Deviations to the regulatory binding documents would require a final statement and or adjustment to the chapter 9.6 and 10.2 in the clinical trial protocol [1], if applicable. Furthermore, the corresponding update of a quality plan should be referenced, which would allow the traceability of the changes occurred during the course of the clinical trial, if applicable [33].

Note to the proposed protocol deviation categorisation:
An informative protocol deviation could be, for example, consistently not reporting any (serious) adverse event for laboratory values, described in the clinical trial protocol as being of interest for the treatment safety profile. The assumed impact to the safety aspects, but also the quality of the reporting, would trigger the assignment to the three different categories. In overall, attention should be drawn to avoiding a clinical trial site and or country induced bias, limiting the acceptance of the data. Furthermore, the correct assignment and documentation of the identified deviations, serious breaches and or issues should allow the early risk-based trend analyses [16], [20], [21], [26].

The presented examples for the items 1-6 outline the overall aspects to be considered, whereas the description for the data items and or processes concern the activities and assumed measures.

Example to item 1: All sites actively recruiting in each country should at least be part of the review at two time-points. This should allow the identification of any significant trend in the quality of the clinical trial site and avoid a systematic error and or critical to quality issue. The experience and suitability of the clinical trial site has to be taken into account. Furthermore, if the clinical trial site was newly selected and or a replacement of the principle investigator would be applicable. Deviations should be recorded as defined in the clinical trial specific quality plan, and classified according to the provided instruction. The completeness of the protocol deviation documentation and adherence to the instructions should be compared with all clinical trial sites. Selection criteria should respect the absence of any deviation, including missing information in the source documentation. This also applies to altered activities, non-conform with the clinical trial plan, and contractual agreement. Deviations due to a safety concern and or potentially resulted because of a gap in the infrastructure of the clinical trial site, are of interest.

Example to item 2: The number of randomised patients is higher or lower as agreed within the clinical trial contractual agreement. Furthermore, if the number of patients falsely included is higher than the assumed average amount due to available comparable information, or no patient was correctly included at all. Discrepancies regarding the assumed average patient number recruited at a clinical trial site, as expected in the defined study population, remains a point of attention. This also concerns the frequency of expected safety events reporting compared to the number of the recruited patients and treatment duration. In addition to that, the amount of included patients with missing assessments, early termination and missing follow-up data.

Example to item 3: The total amount of protocol deviations per clinical trial site and patient is higher or less than the assumed average number of protocol deviations. If applicable, the previous experience and normally occurrence of human errors should be taken into account. The number of similar protocol deviations in each patient at a clinical trial site, as well as the occurrence of the same deviations at more than one site. For the identified protocol deviations, the anticipated impact on the outcome measures and safety for the patient would induce the assignment to the proposed categories (informative, potential informative, non-informative). A summary outcome table and or graphical display could allow an ongoing monitoring regarding the distribution of the observations and assignment to the different categories. This could be supported by descriptive analyses and for presenting the outcome of the review.

Example to item 4: At least two aspects for the selection of the monitoring report should be defined. As proposed in the 2nd edition, the selection of the report could be determined randomly by following an unshaped row (01, 03, 05 and so on) per clinical trial site. Ensuring of one report review for each monitor would be recommended to be scheduled every six months, and one initiation visit report. The second proposed aspect would be deciding in the follow on a risk-based review, depending on the outcome of the previous overall review. First, the quality of the monitoring plan adherence by the monitor, responsible for the clinical trial site, should be paid attention. The aim includes distinguishing between the performance of the monitor and the clinical trial site. Furthermore, to identify any issues potentially inducing, an on-site co-visits assessment. The next aspect concerns the adherence to the clinical trial protocol, as defined in the monitoring plan for the conduct of the clinical trial, by the clinical trial site. The description in the monitoring report should easily allow the retrieving of the clinical trial site procedures. This should include, for example, the collection of the patient informed consent, the management of the adverse event, and if any issues occurred.

Protocol deviations, serious breaches, other deviations and or issues identified by the review for the items 1, 2, 3 and 5, the possible impact on the data for the analyses, the patient safety and or data integrity should be summarised. The overall outcome for the item 4 should present, for example, inconsistencies identified between the quality control activities. Furthermore, the summary should include if a trend and or impact on the overall safety outcome, the primary and secondary clinical trial endpoints, was assumed. Any concerns and or issues expected inducing further assessments, for example, causing financial impact should, and or adjustments to the development program should be summarised briefly in a table [1], [5], [8], [33].

Example to item 5: Clinical trial sites subject to schedule a co-visit or quality assessment visit should be selected according to a pre-defined quality metrics. This is because ensuring a standardised approach over the course of a trial, and if required also on the company level, could increase the quality of the process. This would allow the ongoing monitoring of the risk-based established procedures, for ensuring the data quality and data integrity, as well as the patient safety. An example for a quality metrics with regard to the sponsor oversight was shown in the first edition of the practical guide. Also, the article prepared by D. Whitham et al. [11] provides practical information for the content of a quality metrics. The risk-based decision could also lead to changes in the course of the clinical trial.

Example to item 6: The identified inconsistencies between the for example, clinical trial working instructions, the clinical trial protocol as well as the corresponding contractual agreements should be briefly summarised. Especially, the overall impact on, for example, the compliance with the binding regulatory guidance documents should be justified. It should be described if any corrective and or preventive measures were implemented. Otherwise, the recommendations as outlined in the previous edition remain unchanged. With regard to the pre-study risk score, a brief conclusive summary could be added in an overall table, providing the description and the decision taken. If required, in the follow, the efficiency of the taken measures addressing the pre-study risk would allow the transparent reporting also for any inspection and or audit [1], [16], [26], [27], [28].

Next, the review time-points should be considered, by taking the clinical trial duration into account. The risk-based approach always requires identifying any gaps, discrepancies, misleading data and or processes in a timely manner. Each site, despite only one or few patients were assigned to the clinical trial treatment, should at least two times be part of the data review. Also, the proposed procedure for the overall clinical data review, selecting every other clinical trial site or randomly would be recommended (for example, in an unshaped row 01, 03, 05, 07), at least for the first two review rounds. This should, according to the best practice, most likely outweigh the number of possible performance deviations and or human errors, typically seen by different clinical trial sites and team members. These human errors could occur unintended and or intended before, during and after the initiation of a clinical trial [1], [5], [8], [12]. The outcome of the assessment, based on the six selected data items, will be cumulated and represent the risk-based separate operational review of the multifactorial component score 1. The description can be found in the chapter 3.2, and corresponding Table 5. The details of the example clinical trial, briefly outlined below, correspond unchanged to the 2nd edition.

Example clinical trial: Study duration 24 months; total number of patients 215 at 30 clinical trial sites. This would correspond to an average number of 5,7 patients to be included at each site within 3–6 months and a recruitment rate of 1–2 patients per months/site would be applicable. Assumed average of 2-3 on-site monitoring (0,5 for the initiation, 0,5 for the close out visit, and 2 purely for the monitoring), 1-2 central remote monitoring visits, at each clinical trial site. Per patient, a maximum of 15 regular visits with 7 data items (total of 105 data items) within 2 years are applicable. This applies for all 215 patients in the whole study.

In summary, taking for each patient a maximum of 15 visits with 7 data items (total of 105 data items) within 2 years into account, all protocol deviations, serious breaches, and or deviations and issues are subject to the assessment. Whereas approximately 10 % – 20 % of the other monitored data and or clinical trial site related procedures should be subject to the performance and GCP compliance assessment. Depending on the outcome of the first review round, for example, due to unexpected new information an additional data item could be selected to the next review, as required.

Nevertheless, in total, approximately 75 % – 80 % of all different tasks related to the clinical trial site activities would be recommended to be part of *the separate operational review*. In the following part, the proposed data items, corresponding processes for the risk-based review are outlined. As pointed out before, it depends on the policy and or standardised procedures established in the quality system of the company. The presented different categories of data items are based on the chapter 4 and 5 of the ICH-GCP and correspond to the investigator but also sponsor responsibilities. It was assumed that the accuracy of those selected data items contribute to the robustness of the data-set for the proper analyses, Furthermore, the last item was added ensuring the general administrative activities, deviations could occur, are part of the review as well as the determination of the pre-risk score [1], [5], [8], [26], [27], [28], [33].

Selected Data Items and or Processes:

- **Visits and assessments**: Number of complete missing visits versus the expected number of visits. The number of visits partially incomplete, for example, complete or repeating missing safety measurements such as laboratory values relevant for safety endpoint analyses, including the vital signs, electrocardiogram (ECG), anti-drug antibodies (ADA), outcome measures to patient reported outcomes and quality-of-life assessment (Qol).

- **Efficacy assessments:** Number of complete missing visits versus the expected number of visits. The number of visits partially incomplete, for example, complete or repeating missing efficacy data items. Frequency of non-valid assessments, and or performed with not authorized equipment. It also concerns the use of AI-assisted functionalities (refer to Table 1).

- **Safety reporting:** Number of complete missing serious adverse events (SAE) reports, and number of missing follow-up reports or missing relevant information. Number of reported adverse events (AE), as well as the correct assignment and grading according to the source documentation [34]. The number of suspected, unsuspected serious adverse events (SUSAR) and adverse events of special interest (AESI). The average amount of the informative, potential informative protocol deviations is higher or below the total assumed average amount, anticipated and based on the clinical trial design and clinical trial indication. Adherence to the applicable information provided in the clinical trial protocol, most recent reference safety information, as referenced in the investigators' brochure (IB), and or product information for use of approved medicines (PI) should be a quality criterion.

- **Medical diagnosis:** The number of patients deviating from the in-exclusion criteria and or with a misleading medical diagnosis. For the standardised assignment of the diagnosis, "The International Statistical Classification of Diseases and Related Health Problems" 10th Revision (ICD-10) would be applicable [35].

- **Histology, if applicable:** The number of patients not correctly included into the clinical trial based on the for example, available histology assessment.

- **Baseline, selected study visit documentation, investigator site file:** The number of identified discrepancies by the monitor activities, and the quality of the documentation. The quality and completeness of the investigator site file management with regard to the defined criteria in the contractual agreement and or document management plan.

- **General administration**: Set-up phase, quality system sponsor, including the quality documentation and the trial master file (TMF).

Furthermore, the qualification of cooperation partners, and or external consultants as well as institutions and or participating country, should be part of an overall pre-study risk score assignment. The corresponding pre-defined criteria for the assignment of the risk can be found in the Table 1 "Criteria for the Pre-study Risk Score". The Table 1 below displays the risk category, which indicates the anticipated relevance of the protocol deviation and or serious breaches. It is expected for the preventive identifying of assumed deviations, for example, not in line and or not according to the applicable binding regulatory requirements [1], [5], [8], [12]. The pre-study risk score could induce separate criteria other than presented in the chapter 2.

Table 1: Example criteria for the pre-study risk score (source: own representation) [1], [12], [26], [29].

No.:	Task	Critical 3	Medium 2	Low 1
01	**General administration:** *(Set-up phase, quality system sponsor (also included are the quality documentation and the TMF).*	**Question:** Clinical Trial population specifics, exposed to AI supported products, for example, batteries for the functionality of a medical device. Yes, battery loaded functionality is applicable. No technical documentation for the AI product is present. Manufacturing: Process for the AI and methodical approach not yet known in Europe, Germany. The manufacturing and methodical approach could not be found in the public databases.	**Question:** Clinical Trial population specifics exposed to AI supported products, for example, batteries for the functionality of a medical device. Yes, battery loaded functionality is applicable. The associated risk would according to the limited technical documentation be in a controlled manner. Manufacturing: Process for the AI and methodical approach not yet known in Europe, Germany The manufacturing and methodical approach could be found in the public databases but with less experience.	**Question:** Clinical Trial population specifics exposed to AI supported products, for example, batteries for the functionality of a medical device. Yes, battery loaded functionality is applicable. The associated risk would according to the technical documentation be limited. Manufacturing: Process for the AI and methodical approach known in Europe, Germany but no corresponding validation documents are available. Known, experience present with minimal issue in the technical documentation.
02	**Safety reporting:**	Technical infrastructure 24 hrs/7days/week. Technical limitations/restrictions due to country specific requirements. Restrictions, for example, related to the software provider by the national law. Only locally manufactured and controlled software applications are allowed.	Technical infrastructure 24 hrs/7days/week. Technical limitations/restrictions due to country specific requirements. Restrictions related to the software download by the national law. Software applications of other providers can be used.	Technical infrastructure 24 hrs/7days/week. Technical limitations/restrictions due to country specific requirements. Restrictions related to the availability by the national law (resource issues). Software applications of other providers can be used 24hrs/5 days/week. .
03	**Efficacy assessments:**	**Question:** Routine assessments and technical features are completely different for the efficacy assessment. No exemptions are allowed. Adjustments need to be confirmed by, for example, the responsible institution or committee.	**Question:** Routine assessments and technical features are different for the efficacy assessment. Exemptions are allowed with fully reimbursement. The data will be shared with the responsible institution or committee.	**Question:** Routine assessments and technical features are partially different for the efficacy assessment. Adjustments could be discussed case by case but need to be confirmed by, for example, the responsible institution or committee.
04	**Medical diagnosis:**	**Question:** The disease classification system was not adjusted to, for example, local specific requirements.. The medical diagnosis will be determined in routine care based on, for example, AI methods, only. No histological chemistry assessments are part of the routine care diagnostic program.	**Question:** The disease classification system was adjusted to, for example, local specifics requirements. The medical diagnosis will be determined in routine care based on, for example, AI methods, only. Also histological-chemistry assessments are part of the routine care diagnostic program.	**Question:** The disease classification system includes, for example, local specific requirements due to genetic profile. The medical diagnosis will be determined in routine care based on, for example, AI, only. Also, histological-chemistry assessments are part of in the routine care diagnostic (previous test runs were quality controlled and validated).
05	**Business model:**	**Question:** > 3 leading business (e.g. other than pharmaceutical development, for example, automobile production, dance studio, involved in the cooperation without agreement and or financial supplemental).	**Question:** > 1 leading business (e.g. other than pharmaceutical development, for example, automobile production, dance studio involved in the cooperation without agreement and or financial supplemental).	**Question:** Max of 1 leading business (e.g. other than pharmaceutical development, for example, automobile production dance studio, involved in the cooperation with agreement and or financial supplemental).
xx	***Other, as relevant assumed risks:*** *(For example, cloud-based archiving and validation processes).*	***Question:*** *Cloud-based AI supported communication and archiving applies. No documentation for validation procedures is present.*	***Question:*** *Cloud-based AI supported communication and archiving applies. Limited documentation for validation procedures is present and software security measures.*	***Question:*** *Cloud-based AI supported communication and archiving applies. Limited documentation for validation procedures is present and a two-step based software security measure.*

The **Table 1** presents five examples related to the clinical trial activities general administrative, safety reporting, efficacy assessment, medical diagnosis, and business model. For the assignment to the different categories for the determination of the assumed risk, critical, medium and mild, a description will be provided. This will be shown in the second, third and fourth tab. The last row in the Table 1 outlines another example, potentially relevant, for the assessment of the pre-study risk score. The risk matrix applicable for determination of the pre-study risk score should be 4–6 described in the quality manual of the applicable institution.

Based on this pre-study risk score, further investigations and or adjustments could be required for the project set-up and or clinical trial design. The outcome of the assessment could also concern the selection of the cooperation partners, affiliates, and so on. This initial pre-study risk score assessment should in the best case be performed within approximately 4–6 weeks. The same categorisation, as previously provided, would remain applicable and could be adjusted as required. As pointed out, the pre-study risk score assessment should consider the business model. In particular, the clinical trial cooperation partners based in a country outside the EU could offer services of interest in the field of information technology. Therefore, the alignment with the applicable Regulation (EU) No 536/2014 and Regulation (EU) 2024/1689 for the use of AI-assisted applications should be confirmed in advance [1], [29].

As outlined before, approximately 10 % – 20 % of the monitored and not yet on-site and or remote monitored visits at each site would be proposed to be subject to the review. Whereas, the protocol deviations and or serious breaches, reported by the monitor, should always be part of a fully second review. Especially, this would contribute, ensuring the accuracy and effectiveness of the remote and on-site monitoring activities as well.

In comparison to the clinical data review, the multifactorial score 1 includes a further review (10 % – 20 %.) related to the occurrence of deviations. It should include already monitored data items and the performance quality of the clinical trial site from either on-site and or remote. The standardised risk-based procedure should be outlined in the corresponding study plan and or, for example, guiding quality plan.

In addition to that, also for the "Risk Based Operational Data Review" a pre-programmed metrics would allow a traceable documentation, including a standardised graphical display. An extract of the relevant review outcome could be presented in the quality assessment report. The Table 5 was updated and provides an example for the categorisation of the multifactorial components score 1 as well as the proposed further divided score 1 [18], [19]. Especially, for the filing in the TMF a final track change version with a change history would be recommended [1], [20], [21], [22], [23].

To summarise briefly: *The rationale for the proposed approach should improve the quality of the clinical outcome data, the performance quality and overall GCP compliance by taking the following consideration into account: The "Risk Based Operational Data Review", as part of the clinical data review, also considers a pre-study risk score assignment. The separate conduct of the operational review, based on the selected criteria, clinical trial design, and assessments, should enhance the reliability of the outcome data and acceptance of the data by regulatory agencies. The proposed quantitative measurement, calculated by pre-defined risk-based criteria, should avoid the frequency of misleading data for the quality assessments. The assignment of the protocol deviations, serious breaches and or other deviations and issues to the different defined categories should allow implementing early on, for example, corrective and preventive measures. The assumed further divided weighted score value, assigned to the multifactorial component 1, could support the reliability of the calculated GCP compliance due to the quantified approach [18], [19]. Furthermore, the applicable written evidence would be available, as required by the regulatory requirements [1], [8]. The overall procedure could potentially be supported by the partial use of validated AI-assisted applications [24], [25],[29]. Especially, the pre-study risk score should enable the prevention of critical deviations. Further investments for research cooperations could possibly appropriately be placed by taking into account the pre-study risk score assessment of the assumed service providers.*

3. Definition of the Risk-Based Operational Data Review

The regulatory requirements do not provide a unique definition for the "Risk Based Operational Data Review", as part of the overall clinical data review. Despite this, in overall the applicable regulatory binding document Regulation (EU) No 536/2014 [1], provides the information related to the risk-based quality management. It became effective by the end of January 2022 with an assumed transition period of three years. The chapter VIII and the articles 47 to 59 are relevant for all participating member states concerning these aspects. The regulatory document refers to the principles of the GCP and therefore to the ICH guidelines. In the 2nd edition of the practical guide, the general definition, as understood and aligned to the ICH-GCP [8], for the quality control tool clinical data review was outlined. The chapter 5 defines the sponsor responsibilities but also the chapters 4, 6 and 8 need to be taken into account, as references for the identification of protocol deviations, serious breaches, other deviations and or quality related issues. Furthermore, the appendix III of provides an overall guidance for the safety reporting. For reporting purposes, it could be of advantage considering the appendix IV as a standard table of content and harmonising the overall reporting and documentation procedure.

As outlined before, the interpretation of the underlying regulatory requirements needs to be assigned and linked to the selected processes and or procedures. For this task, other references as already pointed out in the chapter 1 and 2 were also taken into account.

In the chapter IX, the applicable requirements related to the manufacturing and import of an IMP, including auxiliary medicinal products, are outlined. As noted before, the practical guide does not cover in detail the GMP requirements. Nonetheless, the availability of the corresponding IMP documentation at the clinical trial site and or pharmacy but also at the sponsor organisation remains subject to the quality control measures. Missing and or inconsistent traceability in this documentation should be subject to the separate operational review as well [5], [8], [20], [21], [22], [23].

Especially, if an event resulted in a safety issue for the patient and or could limit the data integrity, further risk-based investigations should be considered proactively. Therefore, the interpretation of the regulatory requirements would induce for the *separate operational review* the implementation of an appropriate deviation management tool. The proactive approach could allow limiting the occurrence of repeating human errors and other reporting issues.

Supportive to the chapter 2 the Table 1 provides an overview of the data and or administrative processes assumed as relevant for the determination of the pre-study risk score. The information and proposed procedure for the risk score calculation with regard to the "Risk Based Operational Data Review" will be outlined in more detail in the chapter 5. Another aspect concerns the procedure for the assignment of a threshold, as well as the justification for an upper level of acceptance related to missing data and or values.

In addition to the regulatory guidance documents the reflection paper risk-based quality management in clinical trials, prepared by the EMA, provides in the chapter 5.2 an explanation for the definition of quality tolerance limits [26]. The most recent and current state of medical and statistical knowledge about the variables to be analysed, as well as the statistical design of the trial, should be considered for the assessment. Results of investigations outside the tolerance range as defined in the applicable clinical trial protocol would therefore be subject to the *separate operational review.* The introduction of a tolerance range and or upper limit for specific clinical trial data parameters would allow a better understanding of severe outliers and or systematic errors. Especially those, identified early on with a possible impact on the patient safety and or limited data quality, could be determined. Furthermore, the EMA named the three sources, trial data, trial protocol procedures, GCP and trial management procedures, for the assignment of the deviations (Table 3). Also, the reflection paper regarding common flaw in managing protocol deviations, prepared by the MHRA, does not specify the limits for the determination of a threshold and or upper limit of deviations. It remains with the sponsor deciding on the appropriate determination and procedural approach [16], [20].

The article published by Getz "*Quantifying Protocol Deviation Experience by Clinical Phase*" notes that no awareness to evident data is available, which would allow an overall quantifying and benchmarking of protocol deviations. An assigned working group of 20 major and mid-sized pharmaceutical companies and contract research organizations (CROs) performed a deviations associated study [10]. Ensuring the alignment to the underlying regulatory requirements as well as the assumed procedure should be kept efficiently. For the assignment to the categorisation and score an example for a pre-defined legend, allowing a quantitative measurement, will be shown in the chapter five. The selection of the appropriate category for the identified protocol deviations and or serious breaches should ensure an objective review and robust outcome data-set.

Even though, as initially pointed out for the sponsor oversight, the aim remains to link the activities to the corresponding items of the ICH-GCP, and other recommendation papers ensuring the alignment to the underlying regulatory requirements [1], [5], [8].

3.1 Risk-Based Operational Data Review

The Table 2 provides the different categories of the applicable ICH-GCP items subject to the overall clinical data review. The separate items are referenced by the binding ICH-GCP chapter 5.2 [8]. The outer tab was added naming the articles and appendices of the Regulation (EU) No 536/2014, with regard to the specific activities protocol deviations, serious breaches, other deviations and or issues could occur. The deviations and serious breaches should be considered at a clinical trial site (ICH-GCP chapter 4, 4.5 until 4.9), the sponsor organisation (ICH-GCP chapter 5), a service provider and or any other cooperation partner, including the department quality assurance level [1]. Therefore, the additional information provided in the Table 2 would be categorised as subject to the risk-based separate operational review.

Table 2: Categorisation of the chapter 5, and 4 ICH-GCP E6 (R2) into groups, aligned to the applicable Regulation (EU) No. 536/2014 (source: own representation) [1], [8].

Category	ICH-GCP (R2) Chapter 5	ICH-GCP (R2) Chapter 4, 4.5 – 4.9; 4,11	Regulation (EU) No. 536/2014
Quality assurance and quality control:	5.1; 5.19; 5.20.	Internal policy.	The regulation refers to the ICH-GCP.
Medical management:	5.3.	4.5, 4.8.	Articles 47, 49.
Biostatistics:	5.4; 5.23.	4.5.	Article 56.
Trial/Project management:	5.5; 5.6; 5.7.	4.5.	Article 47 – 59.
Site management/Monitoring:	5.18.	4.6, 4.8.	Articles 47- 59.
Data management:	5.5, 5.23.	4.7, 4.9.	Article 56.
Safety management:	5.16; 5.17.	4.11.	Articles 52, 53, 54, 55.
Medical writing:	5.4; 5.22.	Not applicable	EU- Clinical Trials regulation refers to the ICH-GCP.
Medical/Science:	5.3; 5.12; 5.16; 5.17.	Not applicable	The regulation refers to the ICH-GCP.

The Table 2 displays an overview of the different data categories subject to the "Risk Based Operational Data Review". In particular, these are linked and correlated to the applicable regulatory binding requirements, which can be found in the header of each tab. In each row, the corresponding articles and items subject to the "Risk Based Operational Data Review" are noted.

The following Table 3 was prepared for categorising the data items, procedures and occurred protocol deviations to the corresponding GCP items. The categorisation, outlined by the EMA as shown in the corresponding reflection paper, was considered [26].

Table 3: Assignment of the data items, provided in the Regulation (EU) No 536/2014 to the ICH-GCP chapter 4, and the reflexion paper for the risk-based quality management (source: own representation) [1], [8, p. Ch. 4], [26].

Number	The Clinical trials Regulation EU No 536/2014 [1]	ICH-GCP Chapter 4 Investigator Responsibility [8, p. Ch. 4]	Reflexion paper item 5.2 Quality tolerance limit [26]
01	Visits and assessments:	4.5 Compliance with Protocol. 4.7 Randomization Procedures and Unblinding. 4.6 Investigational Product (s).	Trial data. Trial protocol procedures and GCP. Trial management procedures.
02	Efficacy assessments:	4.5 Compliance with Protocol.	Trial data. Trial protocol procedures and GCP. Trial management procedures.
03	Safety Reporting:	4.11 Safety Reporting.	Trial data. Trial protocol procedures and GCP. Trial management procedures.
04	Medical Diagnosis:	4.5 Compliance with Protocol.	Trial data. Trial protocol procedures and GCP. Trial management procedures.
05	Histology, if applicable:	4.5 Compliance with Protocol.	Trial data. Trial protocol procedures and GCP. Trial management procedures.
06	Baseline and selected study visit documentation:	4.9 Records and Reports. 4.8 Informed Consent of Trial Subjects. 4.6 Investigational Product(s). 4.5 Compliance with Protocol.	Trial data. Trial protocol procedures and GCP. Trial management procedures.
07	General administration, set-up phase, quality system sponsor (also includes the quality documentation and TMF).	4.1 Investigator's Qualifications and Agreements. 4.2 Adequate Resources. 4.3 Medical Care of Trial Subjects.	Applicable Clinical Trial Protocol. Guideline for good clinical practice E6(R2) 5.0 Quality management. 5.1 Quality assurance and quality control. 5.2 Contract Research Organization (CRO). Reflection paper on risk-based quality management in clinical trials Institution specific quality procedures and quality plan.

The **Table 3** displays the selected data items subject to the "Risk Based Operational Data Review" [1]. In the left outer row of the table, the different selected data items and processes as described in the chapter two are outlined. In the second tab the selected data items of the ICH-GCP chapter 4, investigator responsibilities [8, p. Ch. 4] were assigned. The third tab of the Table 3 references the different categories assumed as relevant concerning the quality tolerance limits [8, p. Ch. 5.2]. This information can be found in the reflection paper risk-based quality management prepared by the EMA. For all categories the deviation related to the trial data, trial protocol procedures and GCP and trial management procedures should be reported [26]. The item 7 concerns the proposed pre-study risk score determination. More details regarding the pre-study risk score can be found in the chapter 5 of the current practical guide, and the examples as described in the Table 1.

As pointed before, no specified threshold and or upper level of acceptance, with regard to missing values, any kind of deviations were referenced in the corresponding recommendations, nor the superior regulatory guiding documents. Nevertheless, as shown in the Table 2, for the categories quality assurance and quality control, medical writing and medical science no corresponding ICH-GCP items for the chapter 4 are assigned. The establishment of the quality assurance and quality control at the clinical trial site depends on the internal quality standards of the institution. According to the regulatory requirements the infrastructure will initially be assessed during a pre-study visit and or later on, as required, based on the quality of the performance [1], [8]. The same applies for the disciplines medical and science, as these aspects are the pre-conditions for conducting a clinical trial with a selected investigator. Also, for example, the need to establish a medical writer at a clinical trial site is up to the institution and their internal scientific processes. Even though, the assigned responsibilities expected for the investigator need to be taken into account for the *separate operational review* and correspond to the ICH-GCP items 4.5 until 4.9, 4.11 (refer to Table 3).

It should be mentioned that increasingly the medicinal institutions develop further in-house competencies for ensuring an appropriate internal infrastructure, also for economical purposes. If the institution takes over the independent role as a sponsor for an academic clinical trial, the obligation for the sponsor oversight activities needs to be ensured.

The remaining investigator responsibilities, according to the ICH-GCP chapter 4.0 – 4.4 and 4.10, are part of the regular sponsor oversight activities. It needs to be a written evidence in place confirming these aspects are in line with the expected requirements. This concerns, for example, the qualification of the principle investigator, the clinical trial staff as well as the presence of a suitable infrastructure.

With regard to the definition of an appropriate threshold and or upper level of acceptance for missing values, and or protocol deviations the following investigation was considered. The SISAQOL missing data working group, developed recommendations to facilitate standard approaches for the PRO analyses in cancer randomised clinical trials (RCTs). The corresponding investigation, assessed as reasonable for to be taken into account to determine a threshold and or upper level of acceptance for other deviations as well, can be found in the article: *"International standards for the analysis of quality-of-life and patient-reported outcome endpoints in cancer randomised controlled trials: recommendations of the SISAQOL Consortium"*, published by C.Coens et al. [17]. The simulation study showed that the effect of missing data rates on PRO findings depends on the type of missing data, for example, informative or non-informative missing data. It was recommended that collecting reasons for missing data is key in assessing the effect of missing data for PRO findings (recommended statement 24 in Figure 2). Further details could be found in the article as referenced [17].

Furthermore, the *"Guideline on Missing Data in Confirmatory Clinical Trials"* provided by the Committee for Medicinal Products for Human Use (CHMP) states in chapter 5.1 [12] no rule is present regarding the maximum number of missing values that would still be acceptable. Within the chapter 6, Handling of Missing Data, the different ways to proceed are shown.

Hence, the aim of identifying early on a trend in missing informative values should with this approach be ensured. Besides the missing assessments, also within the process of the "Risk Based Operational Data Review", gaps related to the used information technology system or infrastructure, appropriately education of the study team, likely to induce systematic errors, may also be early on identified. Therefore, these procedural systems and or conditions should be part of the pre-study risk score assessment (refer to Table 1). Also for the *separate operational review* an ongoing quality monitoring could it be of interest. The article, *"A Second Chance to Get Causal Inference Right: A Classification of Data Science Task"* [31] provides insight to the definition and goal of data science, According to the principle of data science, parallel and retrospective analyses could offer a database for academic institutions and or small-mid-sized companies for taking decisions. This could include deciding to continue cooperation with a clinical trial site and or country. Taking such analyses into account, the long-term monitoring of the different quality tools and outcome data, could offer a scientific sound rationale [1], [33].

3.2 Schematic Overview Operational Risk-Based Data Review Flow

The schematic overview of the clinical data review flow (2nd edition chapter 3.2) was adjusted, indicating the additional information with regard to the "Risk Based Operational Data Review". The revised information can be found in red colour and displays the divided weighted value for the *separate operational review* (refer to Table 5).

4.   Risk-Based Operational Data Review

The "Risk Based Operational Data Review" is part of the clinical trial site data review and includes, as already defined before, the categories as outlined below. Another aspect which was newly added concerns the pre-study risk score assessment. The information and procedure was already provided in the chapter 2 and 3. For consistency, the general information related to the items 4a-4d will be shown.

a)   Review and assessment of protocol deviations (ICH-GCP 5.20.1)

The details outlined in the 2nd edition, chapter 4a, of the practical guide were not required to be updated. In the chapter 6 a proposal for the language and content for presenting the review outcome, as part of the regular internal reporting and quality report, was described.

b)   Monitoring visit report review (ICH-GCP 5.18.6)

The details outlined in the 2nd edition, chapter 4b, of the practical guide were not required to be updated. In the chapter 6 a proposal for the language and content for presenting the review outcome, as part of the regular internal reporting and quality report, was described.

c)   Co-monitoring visits (ICH-GCP 5.18.3)

The details outlined in the 2nd edition, chapter 4c, of the practical guide were not required to be updated. In the chapter 6 a proposal for the language and content for presenting the review outcome, as part of the regular internal reporting and quality report, was described.

d)   Medical management review (ICH-GCP 5.3)

The details outlined in the 2nd edition, chapter 4d, of the practical guide were not required to be updated. In the chapter 6 a proposal for the language and content for presenting the review outcome, as part of the regular internal reporting and quality report, was described.

As referenced before, protocol deviations, serious breaches, other deviations and or issues could also occur during the set-up phase of a clinical trial. Therefore, the pre-study risk score assessment should be established, under other, allowing a robust assumption of the assumed average amount of deviations per patient and per clinical trial site (refer to Table 1). The underlying binding regulatory requirements remain as the binding source for a regulatory inspection and or quality assurance qualification procedure or on-study audit [1], [2], [3], [4], [8], [30].

As noted, the retrieved further publication, *"Development of a standardised set of metrics for monitoring site performance in multicentre randomised trials: a Delphi study"* [11], provides in the Figure 2 an example for the format of a risk-based quality site performance metrics. In addition to that, the limitations with regard to the protocol deviations, serious breaches and or other deviations management and documentation announced by the MHRA, should be addressed with the assumed procedure [26].

The overall approach was already described in the previous chapter 2 and 3. The quality tools a, b and c should ensure the identification of protocol deviations, serious breaches, other deviations and or issues with regard to the multifactorial component score 1. It was decided keeping the medical data review as the fourth aspect for the clinical data review separate. The aim would be ensuring a two-step procedure in the risk-based separate operational review. Any discrepancies identified would require a proper and traceable documentation. This should also include the measures taken, including the corresponding communication, if applicable.

The use of pre-programmed data exports, for example prepared by AI-assisted applications, could be considered [24], [25], [29]. In this case, the content of the medical management review could focus on the review with automatically generated outcome information. It should include defined data-sets to be compared by performing the manual review, at least for specified validation procedures. The quality tools, including the assumed risks, should be described in the corresponding monitoring plan and the overall guiding quality document on the company and or clinical trial level. Furthermore, this should also be considered for the validation procedures potentially applicable for AI-assisted quality tools. The described approach would therefore also address and ensure the traceability, management as well as the documentation of the concerned data items subject to the review [1], [8]. The proposed two-step procedure would allow the analysing and comparison of the outcome data over the course of several clinical trials. This should enable the definition of an overall average threshold as well as upper level of acceptance for deviating from the pre-defined procedures, also on the company level [16], [20], [26].

5.   Risk Score Calculation

As outlined above, in addition to the risk score calculation, the determination of the pre-study risk score would be recommended. In the following part, further details are provided with practical examples. The pre-study risk score could contribute, for example, to the selection and qualification procedure of appropriate business cooperations partners. Furthermore, the preventive identification of potential limiting factors, regarding long-term investments in clinical development programs, could be of interest. This would include the manufacturing of an IMP but also the selection of the countries, and clinical trial sites as well as service providers. It should also enable deciding on appropriate clinical trial designs, and indications for addressing the most important medical need [1], [5], [6], [8].

Therefore, the revised EU-Commission AI Act Regulation (EU) 2024/1689, published on 01 August 2024, [29] was taken into account. It demands the categorisation of AI-assisted functionalities, including AI supported software programs, into the defined for high-risk groups. The regulation determined the national reporting of possible AI induced incidences for an ongoing monitoring. Articles provided by Thompson [24], and Askin et al. [25] point out how AI could change and or challenge the clinical research. Because the proposed risk-based quality procedure could possibly be partially conducted by pre-programmed AI mechanism, the pre-study score would be seen as an appropriate preventive risk-based quality measure. An impact on the sponsor oversight activities and quality risk management in all clinical trials, because of the AI induced challenges, could apply. As noted above, the FDA addressed as well [30] the importance of the validation procedures.

The previously presented 5 Point Likert Scale [9] was classified as a suitable tool and would also be recommended for the assessment of the pre-study risk score.

In addition to that, and as described in the chapter two and three, the assigned weighted value for the "Risk Based Operational Data Review" was further divided. The assumed effect would include addressing specifically the dynamic risks, and the different categories of the deviations [18], [19]. Below, the legend for the assessment of the pre-study risk score provides the categorisation to determine the threshold and or upper level of acceptance. The proposed assessment should cover further supportive categories of data items, in addition to the clinical trial endpoints, safety aspects and essential documents.

Attention should be paid to AI-assisted applications commonly established for the data storage and the management of the IMP. The applicable data protection rules should be considered if a cloud based environment would apply. The pre-study risk score should therefore be determined, for example, with a questionnaire before the conduct of a qualification audit [26], [27], [28], [30].

As shown in the Table 5, for the calculation the protocol deviation score the assigned 15 % was further divided proposed as, informative (10 %), potential informative (3 %), versus non-informative (2 %, but this would depend on several factors and the underlying rationale for the assignment of the weighted value). The adjusted information related to the multifactorial component score 1 can be found in red colour. The legend for the score assessment was kept as before and will be shown for the items 4a, b, c and d, The rationale for the proposed and selected risk score assessment should be part of an internal company policy and or risk-based quality management standard procedure, at the responsible institution [5], [8], [26], [27], [28].

In the previous example provided by IMB, a provider of information solutions [19] for dynamic risks, such as performance, the assessment score is based only on the pre-defined settings. The current impact is not part of the performed assessment. Whereas, the overall risk assessment score, for the complete assessments (named as, for example test case, or test suite in the IT environment) is the average value of the risk assessment score for each row, weighted by the importance. The range for the importance can be set from 1 to 5, and 3 or neutral being the default value. Example: Likelihood = Medium, Impact = Medium, Current Impact = High $(2 + 2 + 3)/3) * 2-1) = 3.666))$.

Example legend for scoring related to the items 4a, b, c and d, respectively of the protocol deviations, serious breaches and or other compliance issues identified:

1. *General adherence to the specific requirements as defined in the corresponding manual and defined procedure.*
2. *The identification of any trends in the occurrence of the relevant, as informative and or potential informative classified observations or those observations seen at different clinical trial sites.*
3. *The number of any observation assessed as likely having at least a possible impact on the outcome measures and or safety endpoints.*

Score 1 to 5: 1 would be the poorest outcome and 5 the best one.

1 = no adherence at all, more than 2 trends observed, related to the clinical trial endpoints, safety aspects including the essential documents, more than two observations with possible impact to the outcome measures and or safety endpoints.

2 = in 3–5 aspects no adherence identified, 2 trends observed, related to the clinical trial endpoints, safety including the essential documents, two observations with possible impact to the outcome measures and or safety endpoints.

3 = in 2 aspects no adherence identified, at least 1 trend observed, related to the clinical trial endpoints, safety including the essential documents, one observation with possible impact to the outcome measures and or safety endpoints.

4 = in 1 aspect no adherence identified, at least 1 trend observed, possibly related to the clinical trial endpoints, safety including the essential documents, at least one observation with possible impact to the outcome measures and or safety endpoints.

5 = in no aspects none adherence identified, no trends observed, related to the clinical trial endpoints, safety including the essential documents, no observations with possible impact to the outcome measures and or safety endpoints.

In the very best case, a compliance of 100 % would be applicable by counting three review time-points as follows: 4 x 5 = 20/4 = 5 * 3 = 15 = 100 %.

Example legend for the pre-study risk scoring

1. *The general adherence to the specific requirements as defined by regulatory agency and defined procedures, including the business model of any cooperation partners, are as expected (low risk).*

2. *The general adherence to the specific requirements as defined by regulatory agency and defined procedures, including the business model of any cooperation partners, are not as expected. The impact of the assumed risks seems acceptable, but should be presented to an authority as a for example, scientific advice, for further recommendations (medium risk).*

3. *The general adherence to the specific requirements as defined by regulatory agency and defined procedures, including the business model of any cooperation partners, are not as expected. The impact of the assumed risks could not be accepted by, for example, an internal board member and or quality specialist assigned for the assessment (critical risk).*

Score 1 to 5: 1 would be the poorest outcome and 5 the best outcome score.

1 = no adherence at all, more than 2 critical observed gaps, related to the assigned categories, clinical trial endpoints, safety aspects including the essential documents, more than two observations with possible impact to the outcome measures and or safety endpoints. *

2 = in 3–5 aspects no adherence identified, 2 trends observed, related to the endpoints, safety including the essential documents, two observations with possible impact to the outcome measures and or safety endpoints. Also, if this would be assumed due to the used systems and or software,

3 = in 2 aspects no adherence identified, at least 1 trend observed, related to the clinical trial endpoints, safety including the essential documents, one observation with possible impact to the outcome measures and or safety endpoints. *

4 = in 1 aspect no adherence identified, at least 1 trend observed, possibly related to the clinical trial endpoints, safety including the essential documents, at least one observation with possible impact to the outcome measures and or safety endpoints.*

5 = in no aspects none adherence identified, no trends observed, related to the clinical trial endpoints, safety including the essential documents, no observations with possible impact to the outcome measures and or safety endpoints.*

In the very best case, a compliance of 100 % would be applicable by counting three review time-points as follows: 4 x 5 = 20/4 = 5 * 3 = 15 = 100 %.

*This would also apply if an AI- assisted application and or software program could have caused the identified deviation. Especially, if an AI-assisted application will be used, the identified risks are required to be monitored on an ongoing basis. A comparison table could outline the occurrence of the anticipated risks identified during the pre-study score assessment. The documentation should preferably be stored in a separate table for the assumed on-going validation of the AI-assisted procedures [29], [31].

6. Examples for the Reporting of the Risk-Based Clinical Data Review

The chapter 6 provides a few examples for the description of the review outcome with regard to the regular internal quality reporting. The information provided was based on the example clinical trial, briefly summarised below. As noted before, these are randomly compiled examples and do not reflect one single clinical trial due to confidentiality and data protection requirements. In case of similar descriptions to others, this could be caused by default. Nonetheless, the demand for transparency in the reporting of the results by the ICH-GCP, also referencing to the *Declaration of Helsinki,* could outline the one or other similar case [1], [8]. Furthermore, if the outcome represents negative results and or could possibly be rejected by the competent authorities, the results and outcome data are required to be made available for the public within a defined timeframe [33]. The Table 4 provides an example for the assignment of the three proposed categories of protocol deviations.

Example clinical trial: : Study duration 24 months; total number of patients 215 at 30 clinical trial sites. This would correspond to an average number of patients 5,7 patients to be included at each site within 3–6 months and a recruitment rate of 1–2 patients per months/site. Assumed average of 2-3 on-site monitoring (0,5 for the initiation, 0,5 for the close out visit, and 2 purely for the monitoring), 1-2 central remote monitoring visits, at each clinical trial site. Per patient, a maximum of 15 regular visits with 7 data items (total of 105 data items) within 2 years are applicable. This applies for all 215 patients in the whole study.

a) Review and assessment of protocol deviations (ICH-GCP 5.20.1)

The review and assessment outcome could be presented in tables, figures and graphs. The measurements should occur quantitatively, and if required qualitative descriptions. Most importantly, potential outliers with regard to the safety and efficacy parameters should be paid attention. Also for this purpose, the box-plot seems a suitable graph, especially for indicating any trend developments over the time. In overall, for a trend analyses, the displays could be separated between the overall assessment outcome, and per single clinical trial site. Furthermore, it should be recommended aligning the description and graphical display to the corresponding guidance, applicable for the CSR [33]. The display should present the protocol deviations, serious breaches, other deviations and or issues. Below, examples for the review outcome summary reporting can be found in more detail. In addition to that, it could be of benefit preparing a separate table for the tracking of the review time-points, and corresponding relevant outcome, for further risk-based decisions, if deemed as appropriate.

Example text for the different categories of the multifactorial component:

Assessment 1: Duration of the review time 01.01.2023 until 30.06.2023 (6 months).

In total, at 17 of 30 (~ 57 %) clinical trial sites within the first 6 months 72 protocol deviations were identified in 62/78 (~ 79.5 %) patients in both countries. Thereof, 14/ 72 (19.4 %) were identified in 32/32 (100 %) patients at 7/10 (~70 %) recruiting sites in Austria. The remaining ones were identified in 28/30 (~ 93 %) patients at 8 /16 (50 %) recruiting sites in Germany. In total, four sites in Germany did not yet recruit any patient so far. The identified protocol deviations at the clinical trial sites, located in Austria were all classified as non-informative by the monitor and categorised to the following categories: 5/14 (~ 35 %) partially missing laboratory assessments, at mostly three different assessment visits, thereof in 2/5 (40 %) cases the missing values were seen at the baseline visit. In all cases, the relevant value to assess the primary endpoint were missed. All clinical sites were reminded to investigate the underlying reason and to ensure the proper availability of the value, especially the baseline value, required for the determination of any change occurred over the time. The remaining 9/14 (~ 64 %) protocol deviations belonged to the category safety reporting. In 5/9 (~ 55 %) cases, the follow-up report of the serious adverse event was not yet send. The five patients at five different clinical trial sites were discharged on the same day and not hospitalised over the night. This information needed up to 3 months to be retrieved, and therefore be categorised as a gap in the communication and documentation. It should be noted that also the safety management group did not request any further follow-up information to the above safety reports. For 4/9 (44 %) other identified protocol deviations, 4/9 (44 %) the identification of the patient identifiers required to raise queries. Thereof, in 3/9 (~33 %) cases the question if the criteria for a serious breach was fulfilled was not addressed. The 32 protocol deviations occurred at seven clinical trial sites located in Germany in 28 patients. All identified protocol deviations occurred at patient visits, expected to administer the investigational medicinal product. In 14/32 (47.3 %) cases, the label in the pharmacy could not be verified for 18 patients The remaining 28 protocol deviations occurred at the same clinical trial sites in 14 patients. In all cases the vial number, documented in the internal pharmacy management system, did not correspond to the assigned number via the randomisation system. Thus, in 20/28 (~ 71.3 %) cases the correct medication was administered, whereas in 8/28 (14.2 %) cases the vial number and therefore the confirmation of the administered IMP could not be verified.

For the described cases, the same applies as outlined above. With this information, the following concluded outcome could be reported and or induce further investigations: 1. The 8 cases the IMP administration could not be verified should have been considered to be an informative, relevant protocol deviation, despite no safety issue was yet determined. The occurrence was seen in two patients only. 2. The discrepancies identified in the internal documentation system of the pharmacies should be monitored again to avoid any further occurrence. Furthermore, a serious breach was reported preventive. The average amount of the protocol deviations occurred within the 6 months could not be determined.

This is because the responses to the questions raised were not yet send back by the monitor. The information is expected to be provided with the next quality report. In case of a significant change of the patient safety and or data integrity, an interim note should be prepared. Furthermore, the implemented corrective and or preventive actions should be outlined.

The Table 4 provides an example for the assignment of one protocol deviation, identified in more than one case. The defined criteria should indicate the difference between the three categories informative, potential informative and non-informative. The identified protocol deviation refers to the protection of the clinical trial population and was classified as a critical to quality observation.

The next step would be selecting the patients for the monitoring report review be defining risk-based criteria. In overall inconclusive and not resolved, deviations and other issues presented in the monitoring report should induce a co-monitoring visit at the clinical trial site, as required. It depends on the overall occurrence and impact on the patient safety and efficacy outcome data, as well as data integrity and data protection aspects. Furthermore, the decision should take into account the judgement of the assigned project team, according to the internal policy and or quality manual. Below, the example regarding the categorisation criteria will be shown.

Table 4: Example for the categorisation of the protocol deviations (source: own representation) [1], [5], [8], [20], [26].

No.	ICH-GCP	Informative protocol deviation	Potential-Informative protocol deviation	Non-Informative protocol deviation
01	**Protocol deviation: (ICH-GCP 5.18.6).**	Inclusion and exclusion criteria were not met for several patients.	Inclusion and exclusion criteria were not met for several patients but commented by the investigator, with an accepted rationale and proof of written evidence.	Inclusion and exclusion criteria were not met. The sponsor medical assigned person was contacted before the patient consent was signed.
02	**Monitoring visit report review: (ICH-GCP 5.18.6).**	The deviation to the inclusion and exclusion criteria was described for all patients but not traceable assigned to the different patients.	The deviation to the inclusion and exclusion criteria was described but only for three of seven concerned patients.	The deviation to the inclusion and exclusion criteria was described but no remark to the not yet signed patient information consent could be found in the source documents.
03	**Co-monitoring visits: (ICH-GCP 5.18.3).**	The deviation to the inclusion and exclusion criteria could not be found in the source documentation. The assigned quality control person, at the clinical trial site could not provide further information.	The deviation to the inclusion and exclusion criteria could be found for three of seven patients in the source documentation.	The deviation to the inclusion and exclusion criteria could not be found in the source documentation correctly. The applicable cases were documented on the pre-screening log.
04	**Medical management review: (ICH-GCP 5.3).**	The deviation to the inclusion and exclusion criteria was confirmed. The patients safety was not compromised but all patient could not be considered for the efficacy analyses. .	The deviation to the inclusion and exclusion criteria was confirmed. The patients safety was not compromised. For the inclusion in the efficacy analyses the cases were sent to the independent data monitoring group for further advice.	The deviation to the inclusion and exclusion criteria could not be confirmed in the patient listing of the database export. The documentation was not expected to be kept in the database, directly as no patient identification number

				was generated. However, the change history included a query set by the monitor and data management assigned group.

The Table 4 presents one protocol deviation occurred in more than one patient. For each of the four proposed clinical trial related data items, a description of the applicable criteria for the assignment to the category can be found. The data items selected for the example concern the protocol deviations, monitoring visit report review, co-monitoring visits, and medical management, In the first row for the assignment of the identified observations the different categories are provided. It will be distinguished between informative, potential-informative, non-informative protocol deviations.

b) Monitoring visit report review (ICH-GCP 5.18.6)

Assessment 1: Duration of the review time 01.01.2023 until 30.06.2023 (6 months).

In total, at 17 of 30 (~ 57 %) clinical trial sites within the first 6 months, 21 of 60 assumed on-site visits were conducted by 4 monitors (average of 4 visits for each monitor). At 3 of 17 (~ 17 %) clinical trial sites, an additional monitoring visit was required to be scheduled. In overall, no critical but few potential informative observations were identified. The potential major or informative observations were related to the IMP administration. Due to pending responses, the final assessment could not be completed. No safety concern was identified so far, the pending final assessment would not limit the quality of the monitoring activities. Several non-informative observations were seen due to misleading information, but could be resolved without any delay. 14/17 (~ 17 %) of the monitor reports were finalised within the timeframe as determined in the corresponding monitoring plan. The assessment of the open action items could not be performed so far, as mostly one visit was conducted at each active recruiting clinical trial site. It should be noted that nevertheless, the action items and required corrections were closed in 12/17 (~ 81 %) monitoring reports, at the time of the report finalisation. Thus, no major and or informative concern was raised, except the open responses related to the IMP administrations. In overall, no serious breach was identified.

Depending on the clinical trial objectives, further assessments could be added accordingly.

c) Co-monitoring visits (ICH-GCP 5.18.3)

No on-site co-monitoring visit was performed so far. Nevertheless, in total, the assigned sponsor representative attended 20 % of the site initiation visits. The site selection assessment was reviewed again for the clinical trial sites subject to the review. The review set focus on the infrastructure, the laboratory sample storage and shipment processes, as well as the internal information technology systems. In addition to that, the pharmacy was visited as well to briefly review the IMP supply documentation, storage conditions as well as the temperature recording. In 8/30 (~27 %) cases, the description in the site selection report did not correspond to the current performed drug accountability procedure.

Depending on the clinical trial objectives, further assessments could be added accordingly.

d) Medical management review (ICH-GCP 5.3)

In total, for 44/78 (~ 56 %) patients, a request for clarification needed to be answered before the patient was included into the clinical trial and assigned to the treatment arm. During the review period, another 56 clarification requests were sent to decide on questions related to the clinical trial protocol interpretations by 13/30 (~ 43 %) clinical trial sites. These clarification requests were categorised as relevant and informative for the management of the AESI. It was decided to add a notification letter to the next regular newsletter. The current documentation and reporting of the occurred questions and clarifications could be seen as in line with the corresponding guiding medical review and quality plan.

Depending on the clinical trial objectives, further assessments could be added accordingly.

As emphasised, the quality outcome reporting requires the alignment to the regulatory requirements. The most recent recommendations, for example in the follow of an audit and or inspection findings, could be addressed. The outcome reporting should briefly conclude and present the results. As shown and outlined in the regulatory binding guidance documents, any impact on the data integrity, patient safety and or efficacy data should be transparently made available and reported. To remark again, the definition of a specific threshold for the protocol deviations, serious breaches, other deviations and or issues was not explicit mentioned in the guidance documents. However, as concluded in the chapter 2 and 3, the proposed approach would offer the appropriate categorisation of the different deviations. For the transparent reporting also potentially identified and or repeating issues, indicating a potential trend, should be summarised. The level of detail in the reporting depends on the overall quality reporting procedure established by the corresponding research organisation [26].

.

7. Quality Control

The purpose of this chapter is to describe and underline the justification of the established risk-based quality management activities and procedures, within a sponsor and or sponsor representative organisation. It also could serve, for example, supporting the quality management system certification, an internal audit and or regulatory inspection. In addition to the proposed approach for the "Risk Based Operational Data Review" the following further quality control aspects should be paid attention.

The determination of a threshold and or upper level of acceptance for the different protocol deviations, serious breaches, other deviations and or issues should be monitored based on the first baseline review assessment. The lower level of acceptance would not apply in all cases. Attention should be given to a sponsor organisation or clinical trial site without any deviation and, for example, an interim remote quality assessment as a safety measure could be considered. Depending on the assessment outcome, the assumed threshold could be adjusted, if deemed as appropriate. In overall, the rationale for the decision taken should be documented beforehand and, if required, discussed with the quality assurance personnel. This would allow, for example, independent interim reviews for data integrity but also safety reasons [1], [2], [3], [4], [26], [30].

The implementation and use of technical AI-assisted functionalities, allowing a partially pre-programmed review of the data-sets, could be considered. In the last five years, publications related to AI assisted applications could be noted, also in the health care and scientific field. These different applications could possibly require adjusted quality assurance procedures and validation procedures [5], [8], [24], [25], [29].

As noted, the referenced underlying regulatory guidance's, national law and corresponding recommendations do not provide a single standard approach or procedure, for the separate operational review. The quality plan and or any other internal guiding document should provide a standardised approach, also related to the applicable risk-based method, for the performed quality review.

As outlined before, in the underlying rationale for the proposed multifactorial component score 1 the following approach could be considered. The items a) and d) should, according to the understanding of the applicable regulatory binding guidance, be categorised as superior quality tools. Whereas, the items b) and c) would be assigned as supportive quality activities. It should enable identifying the quality of the monitoring activities and inconsistencies between the outcome of the review and documentation procedures [1], [12], [16].

The Table 5 was revised accordingly and outlines the divided value for the assigned 42.5 % in red colour [18], [19]. The procedure would allow the continuous overall assessment ensuring a standardised approach over all clinical trials, conducted by a sponsor and or outsourced activities to a CRO.

In addition to that, the traceable documentation of the complete review procedure including the outcome would ensure the requirements and recommendations with regard to the data integrity published by, for example, in the WHO 5 TRS 996 Annex and by the MHRA [20], [21], [22], [23].

For the continuous oversight of the clinical data, and the separate operational review, the use of an AI assisted application could possibly support the monitoring of the assigned threshold and or upper level of acceptance for deviations. It could possibly require the definition of further appropriate regular validation procedure for ensuring the regulatory requirements [1], [8], [24], [25], [29]. For a validated assessing of the GCP compliance, these procedures could also contribute to confirm the weighted value. As shown in Table 5, it would be counted together and divided with a defined factor. The same was assumed for the on-site verification with regard to the outcome of plausibility checks by remote monitoring activities. In particular, addressing open questions, assessments and conclusions would, according to the understanding of the legally binding regulations, indicate the quality level of the separate operational review [20], [21], [22], [23]. Avoiding ambiguous data and or invalid results, for example, possibly caused by the established procedures and infrastructure of the clinical trial site and or sponsor organisation should then be feasible.

Otherwise, the information provided in the previous edition remains as before. Especially, for open action items, discrepancies not resolved at the time of the regular quality reporting an interim note or report should be prepared [8], [21], [22], [23], [26].

Table 5: Overall GCP clinical trial compliance metrics (source: own representation) [1], [12], [16], [18], [19].

No.:	Category of the Clinical Data Review	Single Score 1st review	Single score 2nd review	Single score 3rd review	Cumulated Score	% weighted for the GCP "clinical data review"	% GCP clinical trial compliance of the "Clinical Data Review" 80 % to 100 % very good; 75 % to 80 % good; < 75 % medium
1.	Clinical Data Review: (a) Score	3	4	2	NA	35%of 42.5 %	35 % of 0.304
	(Protocol deviations, monitoring visit review, co-monitoring visits) [2].	4	5	3	NA	15 % of 42.5 % Informative 10 % of 15 % Potential informative 3 % of 15 % Non-informative 2 % of 5 %	15 % of 0.304
NA[3]	Cumulated score per time-point:	15/20 (75 %)	16/20 (80 %)	12/20 (60 %)	43/60 (71.6 %)	42.5 %	0.716*0.425 = 0.304
2.	Score Safety Data Review:	11/15 (73.3 %)	13/15 (86.6 %)	8/15 (53.3 %)	32/45 (71 %)	42.5 %	0.71*0.452 = 0.302
3.	Score Medical Coding Review:	11/15 (73.3 %)	13/15 (86.6 %)	8/15 (53.3 %)	32/45 (71 %)	15.0 %	0.71*0.425 = 0.106
NA[3]	Total Clinical Data Score:	NA	NA	NA	(71.3 %)	NA	NA
NA[3]	Total GCP Clinical Trial Compliance:	NA	NA	NA	NA	~ 71.2 %	0.304+0.302+0.106 = 0.712 *100 = ~71.2 %

---

[2] The table 3 displays highlighted in red the new additional information with regard to the multifactorial component 1. Examples can be found on the website: https://sciencing.com/weighted-score-10047404.html [18], [19].

[3] Not applicable refers to the assignment of a number in the tab. The rows three, six and seven display the calculate. Therefore, the completion of the fields marked with NA do not contain a value.

a)   **Quality Assessment Report Item Risk-Based Operational Data Review**

Also for this item the information provided in the 2nd edition of the practical guide remains applicable. As outlined before, the different quality tools assigned to the "Risk Based Operational Data Review" could be presented in one chapter by describing the retrieved information separately and or in sub-categories. In overall, the clinical trial site GCP compliance, and the complete clinical data review should be described separately. Both aspects are subject to the determination of the overall GCP clinical trial compliance. The examples provided in the chapter 6 should be summarised with a brief conclusion. In general, the alignment of the activities to further applicable guidance's requested by any third participating member state and or country, should be clarified beforehand. In particular, the safety reporting and or reporting of, for example, important informative serious breaches, also with regard to the manufacturing procedure, could apply [1], [2], [3], [4], [5], [8].

The corresponding summary description of the overall data quality assurance activities would be expected in the chapter 6 of the CSR [33]. Whereas, the summary of the relevant, and or informative and potential informative protocol deviations, including any actions taken or impact on the study population or evaluable patient data, should be described separately, according to the guidance ICH E3. With regard to the displays and the visual presentation of the data in tables, figures and graphs, the tools and examples would remain as the example shown in the 2nd edition of the practical guide. Examples, as provided by the quality expert group ASQ, s were seen as suitable for ensuring transparent graphical displays (https://asq.org/quality-resources/histogram).

8.   Sponsor Oversight Plan – Cross References

To ensure that the study plans will be kept consistent, the implemented review to assess the multifactorial component score 1 should normally be part of the description in the study plans as outlined in the chapter 8 of the previous 2nd edition (Study plans and manuals cross-references; (source: own representation)). In general, specific description and or quality procedures are expected to be part of the overall guiding quality plan on the company but also clinical trial level [1], [5], [8].

9.   Conclusion

In addition to the overall previously provided conclusion regarding the conduct of the sponsor oversight activities, the specific aspects related to the "Risk based Operational Data Review" can be found below. The corresponding regulatory binding framework remains the Regulation (EU) No 536/2014, and other guidance's [1], [2], [3], [4]. The reinforcement of the importance regarding the establishment of the sponsor oversight in all clinical trials was also emphasised by the FDA in April 2023 [30]. The proposed risk-based *separate operational review* should be categorised as part of this requirement. As noted, it was classified as a separate activity and part of the clinical data review. In addition to that, it could enhance the quality of the risk-based quality management procedures and therefore strengthen the sponsor oversight activities.

Implementing partial pre-programmed data exports by pre-defined risk-based variables could possibly increase the quality of the review. As outlined by D. Thompson [24], and S. Askin, et al. [25] AI assisted applications in the field of clinical research could offer interesting economic developments but as well induce challenges. The revision of the AI Act Regulation 2024/1689 provides an information table outlining high-risk AI assisted functionalities [29]. AI-assisted tools, categorised as suitable to be implemented in the operational conduct of clinical trials, are currently related to the recruitment and other tracking activities [24], [25].

As pointed out before, the AMNOG procedure in Germany [6], [7] considers the health status development over the time, and the safety and efficacy data. The *separate operational review* was assessed as a suitable approach for ensuring the robustness and expected quality of the data set for the AMNOG assessment [6]. Ensuring the recommendation, avoiding as much missing data items and protocol deviations as possible, should with the procedure be feasible. According to the best knowledge, the proposed categorisation, distinguishing between informative, potential informative and non-informative deviations, should fulfil the adherence to the regulatory binding guiding documents [1], [8]. Furthermore, the recommendations provided in reflection papers by the EMA, MHRA and other expert groups should be addressed as well [20], [21], [22], [26], [27], [28].
In particular, the proposed categorisation and definition of the tolerable amount of missing data items and or missing visits, published by C. Coens et al. (SISAQOL Consortium), [17] was categorised as a suitable measure. Coens addressed the consistency of the quality-of-life data and categorising missing data. Also, for defining a threshold, and or upper level of acceptance related

to protocol deviations the provided rationale by Coens was seen as a possible methodical risk-based approach [1], [8], [26].

To remark again, despite the assigned investigator is responsible for the conduct of the clinical trial, the data quality and clinical procedures at the clinical trial site, the sponsor remains the responsible party for ensuring the proper oversight activities. In several cases during regulatory inspections and or audits at the clinical trial site, a lack of compliance was also categorised for the sponsor organisation. Therefore, the sponsor oversight activities should include, as well, the "Risk-Based Operational Data Review". The proposed procedure includes the assessment of the clinical trial site processes and performance issues. It should prevent misleading information was potentially provided by the sponsor organisation and not correctly authorized by the assigned sponsor representative [1], [8], [23], [27], [30].

Regarding the assigned value for each presented quality control tool, a further divided value for the multifactorial component score 1 was assessed as an appropriate approach [18], [19]. This additional step should support the overall quality of the risk-based review [26]. In addition to that, it was concluded that the different selected data items and categories assumed for the review should ensure quantitative measurable analyses. For allowing a unique approach, the proposed 5 Point Likert Scale could also be recommended for the assignment to the applicable category [9]. In the chapter 4 the proposed legend and description for the different categories can be found.

Discussions regarding the handling, management of protocol deviations prepared by the *"Transcelerate Group"* were also considered [13], [14], [15] for proposing a pre-study risk score assessment. The determination of a pre-study risk score was categorised as an appropriate measure for the potential changes in business cooperations induced by, for example, AI-assisted functionalities. AI functionalities, applicable for the manufacturing procedure of an investigational medicinal product, could increasingly be expected. The underlying GMP related regulatory requirements including the revised Annex 1, remain the legally binding framework.
Preventive clarifications, and or discussions induced as a result of the pre-study risk score could support the appropriate decisions. Especially, written evidence for the validation processes and the threshold for deviations related to the for example, automated manufacturing process, and management, would require a quality procedure [1], [5], [8].

Investigations related to the occurrence of protocol deviations, and or quality assurance issues in already closed clinical trials, could offer a guiding value for the determination of an average threshold. It was assumed that AI-assisted applications could possibly be established, in the field of data science, and supporting an ongoing monitoring of protocol deviations, other deviations, and or issues assumed as critical to the quality aspects. Further comprehensive systematic reviews and or meta-analyses, by collecting this information as well as trends, could be a relevant investigation [31]. It could offer the possibility of covering the gap outlined by, for example, the MHRA, regarding the transparency of the protocol deviations management [16].

For ongoing internal quality assurance monitoring activities on the company level the investigation, conducted and published by K. Getz, *"Quantifying Protocol Deviation Experience by Clinical Phase"*, offers relevant information [10].

Finally, the overall aim of the practical guide was presenting a possible interpretation of the regulatory binding and supportive guidance documents. As noted, it depends on the size, the business model as well as the outsourcing, risk based quality strategy or other aspects of a sponsor organisation implementing a regulatory accepted approach.

Thus, the rationale as presented in the current edition would, according to the understanding and judgement, ensure the appropriate tools are implemented. This would ensure the delivery of a robust, reliable and stable data-set for the outcome analyses and reporting of the clinical trial results [12], [20], [33]. Nevertheless, it also could depend on the interpretation of an assessing country. The Regulation (EU) No 536/2014 was assumed for the alignment of the assessments of the concerned member states [1].

It was concluded, that a change in the business model, including the outsourcing strategy due to AI assisted applications, and other aspects could not be excluded within the next years [24], [25], [29].
The next point of relevance for further discussion could be related to the current developments regarding the "Industry 4.0 criteria, Manufacturing x" procedures, as also pointed out by the FDA. This is also because of the anticipated international economic developments in the Asia-Pacific Economic Cooperation (APEC) as announced by, for example, the World Trade Organization (WTO).

References

[1] „Clinical trials - Regulation EU No 536/2014 - Public Health - European Commission", Public Health. Zugegriffen: 14. Oktober 2018. [Online]. Verfügbar unter: /health/human-use/clinical-trials/regulation_en

[2] „AMG - Gesetz über den Verkehr mit Arzneimitteln". Zugegriffen: 8. April 2018. [Online]. Verfügbar unter: https://www.gesetze-im-internet.de/amg_1976/BJNR024480976.html

[3] „Clinical trials - Directive 2001/20/EC - Public Health - European Commission", Public Health. Zugegriffen: 14. Oktober 2018. [Online]. Verfügbar unter: /health/human-use/clinical-trials/directive_en

[4] „GCP-V - Verordnung über die Anwendung der Guten Klinischen Praxis bei der Durchführung von klinischen Prüfungen mit Arzneimitteln zur Anwendung am Menschen". Zugegriffen: 19. Mai 2018. [Online]. Verfügbar unter: http://www.gesetze-im-internet.de/gcp-v/BJNR208100004.html

[5] „EudraLex - Volume 4 - European Commission". Zugegriffen: 27. November 2024. [Online]. Verfügbar unter: https://health.ec.europa.eu/medicinal-products/eudralex/eudralex-volume-4_en

[6] „Zusatznutzen neuer Arzneimittel – Kategorien - Gemeinsamer Bundesausschuss". Zugegriffen: 12. Oktober 2022. [Online]. Verfügbar unter: https://www.g-ba.de/themen/arzneimittel/arzneimittel-richtlinie-anlagen/nutzenbewertung-35a/zusatznutzen/

[7] „BfArM - Gesetze und Kostenverordnung". Zugegriffen: 2. April 2023. [Online]. Verfügbar unter: https://www.bfarm.de/DE/Arzneimittel/Klinische-Pruefung/Genehmigungsverfahren/Gesetze-und-Kostenverordnung/_node.html

[8] „Good Clinical Practice (GCP) : ICH -GCP_E6_R2". Zugegriffen: 3. Juni 2018. [Online]. Verfügbar unter: http://www.ich.org/products/guidelines/efficacy/efficacy-single/article/integrated-addendum-good-clinical-practice.html

[9] „5 Point Likert Scale Analysis, Interpretation and Examples". Zugegriffen: 20. September 2022. [Online]. Verfügbar unter: https://ppcexpo.com/blog/5-point-likert-scale-analysis

[10] K. Getz, „Quantifying Protocol Deviation Experience by Clinical Phase", *Applied Clinical Trials*, Bd. 31, Nr. 6, Juni 2022, Zugegriffen: 2. Dezember 2022. [Online]. Verfügbar unter: https://www.appliedclinicaltrialsonline.com/view/quantifying-protocol-deviation-experience-by-clinical-phase

[11] D. Whitham *u. a.*, „Development of a standardised set of metrics for monitoring site performance in multicentre randomised trials: a Delphi study", *Trials*, Bd. 19, Nr. 1, S. 557, Okt. 2018, doi: 10.1186/s13063-018-2940-9.

[12] EMA, „Missing data in confirmatory clinical trials - Scientific guideline", European Medicines Agency. Zugegriffen: 24. März 2023. [Online]. Verfügbar unter: https://www.ema.europa.eu/en/missing-data-confirmatory-clinical-trials-scientific-guideline

[13] „TransCelerate - Protocol Deviations in Clinical Trials", TransCelerate. Zugegriffen: 22. Juli 2022. [Online]. Verfügbar unter: https://www.transceleratebiopharmainc.com/initiatives/protocol-deviations/

[14] „TransCelerate-RBM-Position-Paper-FINAL-30MAY2013.pdf". Zugegriffen: 8. März 2017. [Online]. Verfügbar unter: http://www.transceleratebiopharmainc.com/wp-content/uploads/2013/10/TransCelerate-RBM-Position-Paper-FINAL-30MAY2013.pdf

[15] „TransCelerate-RBM-Update-Volume-I-FINAL-27JAN2014.pdf". Zugegriffen: 16. September 2017. [Online]. Verfügbar unter: http://www.transceleratebiopharmainc.com/wp-content/uploads/2014/01/TransCelerate-RBM-Update-Volume-I-FINAL-27JAN2014.pdf

[16] „MHRA Identifies Common Flaw in Managing Protocol Deviations | Veeva Systems EU". Zugegriffen: 30. November 2022. [Online]. Verfügbar unter: https://www.veeva.com/blog/mhra-identifies-common-flaw-in-managing-protocol-deviations/, https://www.veeva.com/eu/blog/mhra-identifies-common-flaw-in-managing-protocol-deviations/

[17] C. Coens *u. a.*, „International standards for the analysis of quality-of-life and patient-reported outcome endpoints in cancer randomised controlled trials: recommendations of the SISAQOL Consortium", *Lancet Oncol*, Bd. 21, Nr. 2, S. e83–e96, Feb. 2020, doi: 10.1016/S1470-2045(19)30790-9.

[18] „How to Do a Weighted Score", Sciencing. Zugegriffen: 20. Oktober 2022. [Online]. Verfügbar unter: https://sciencing.com/weighted-score-10047404.html

[19] „How risk assessment scores are calculated - IBM Documentation". Zugegriffen: 13. September 2022. [Online]. Verfügbar unter: https://www.ibm.com/docs/en/elm/6.0.5?topic=risk-how-assessment-scores-are-calculated

[20] MHRA-Medicines and Healthcare products Regulatory Agency, „MHRA GxP Data Integrity Definitions and Guidance for Industry - GOV.UK". Zugegriffen: 28. November 2018. [Online]. Verfügbar unter: https://www.gov.uk/government/news/mhra-gxp-data-integrity-definitions-and-guidance-for-industry

[21] „Reflection paper on GCP compliance in relation to trial master files (paper and/or electronic) for management, audit and inspection of clinical trials", S. 17.

[22] Good clinical practice inspector working group, „Guideline on GCP compliance in relation to trial master file (paper and/or electronic) for content, management, archiving, audit and inspection of clinical trials", S. 17, März 2017, [Online]. Verfügbar unter: https://www.ema.europa.eu/documents/scientific-guideline/draft-guideline-good-clinical-practice-compliance-relation-trial-master-file-paper/electronic-content-management-archiving-audit-inspection-clinical-trials_en.pdf

[23] „WHO Annex 5 Guidance on Good Data and Record Management Practices, 2016", Validation Center. Zugegriffen: 3. Oktober 2022. [Online]. Verfügbar unter: https://validationcenter.com/library/library/regulations-and-guidelines/annex-5-guidance-good-data-record-management-practices-2016/

[24]   D. Thompson, „CROs and Artificial Intelligence: How AI is Changing Clinical Research ∗ Vial", Vial. Zuge-
       griffen: 8. Juni 2023. [Online]. Verfügbar unter: https://vial.com/blog/articles/cros-and-artificial-intelligence-
       how-ai-is-changing-clinical-research/

[25]   S. Askin, D. Burkhalter, G. Calado, und S. El Dakrouni, „Artificial Intelligence Applied to clinical trials: oppor-
       tunities and challenges", *Health Technol.*, Bd. 13, Nr. 2, S. 203–213, März 2023, doi: 10.1007/s12553-023-
       00738-2.

[26]   European Medicines Agency, „Reflection-paper-risk-based-quality-management-clinical-trials_en.pdf". Zu-
       gegriffen: 8. Oktober 2018. [Online]. Verfügbar unter: https://www.ema.europa.eu/documents/scientific-gui-
       deline/reflection-paper-risk-based-quality-management-clinical-trials_en.pdf

[27]   EMA, „Good Clinical Practice Inspectors Working Group", European Medicines Agency. Zugegriffen: 2. Ap-
       ril 2023. [Online]. Verfügbar unter: https://www.ema.europa.eu/en/human-regulatory/research-develop-
       ment/compliance/good-clinical-practice/good-clinical-practice-inspectors-working-group

[28]   ICH Expert Working Group, „Quality Risk Management : ICH". Zugegriffen: 14. Oktober 2018. [Online]. Ver-
       fügbar unter: http://www.ich.org/products/guidelines/quality/quality-single/article/quality-risk-manage-
       ment.html

[29]   „AI Act | Shaping Europe's digital future". Zugegriffen: 16. März 2024. [Online]. Verfügbar unter: https://digi-
       tal-strategy.ec.europa.eu/en/policies/regulatory-framework-ai

[30]   „Final FDA Guidance Reinforces the Importance of Sponsor Oversight and Quality Risk Management in All
       Clinical Trials". Zugegriffen: 9. August 2023. [Online]. Verfügbar unter: https://www.sidley.com/en/in-
       sights/newsupdates/2023/04/final-fda-guidance-reinforces-the-importance-of-sponsor-oversight

[31]   M. A. H. John Hsu &Brian Healy, „Full article: A Second Chance to Get Causal Inference Right: A Classifi-
       cation of Data Science Tasks". Zugegriffen: 20. September 2022. [Online]. Verfügbar unter:
       https://www.tandfonline.com/doi/full/10.1080/09332480.2019.1579578

[32]   „ICH E6 (R2) Good clinical practice - Scientific guideline | European Medicines Agency (EMA)". Zugegrif-
       fen: 24. November 2024. [Online]. Verfügbar unter: https://www.ema.europa.eu/en/ich-e6-r2-good-clinical-
       practice-scientific-guideline#revision-3-8264

[33]   EMA, „ICH E3 Structure and content of clinical study reports", European Medicines Agency. Zugegriffen: 4.
       Oktober 2022. [Online]. Verfügbar unter: https://www.ema.europa.eu/en/ich-e3-structure-content-clinical-
       study-reports

[34]   „CTCAE_v5_Quick_Reference_5x7.pdf". Zugegriffen: 17. Oktober 2022. [Online]. Verfügbar unter:
       https://ctep.cancer.gov/protocolDevelopment/electronic_applications/docs/CTCAE_v5_Quick_Refe-
       rence_5x7.pdf

[35]   „Was sind ICD- und OPS-Codes?", gesund.bund.de. Zugegriffen: 5. November 2022. [Online]. Verfügbar
       unter: https://gesund.bund.de/was-sind-icd-und-ops-codes

List of Tables

Abbreviations

ADA Anti-drug Antibodies
AE Adverse Events
AESI Adverse events of special interest
AI Artificial intelligence
Altered term: Separate operational review Risk Based
   Operational Data Review
API Active pharmaceutical ingredient
ASQ Quality expert group

CHMP Committee for Medicinal Products for Human
   Use
CRO Contract research organization
CSR Clinical study report
CTP Clinical trial protocol

ECG Electrocardiogram
EMA European Medicines Agency
EU European Union

FDA Food and Drug Administration

G-BA Gemeinsamer Bundesausschuss
GCP Good Clinical Practice

GDP Good Documentation Practice
GLP Good Laboratory Practice
GMP Good Manufacturing Practice

IB Investigators brochure
ICD-10 International Statistical Classification of
   Diseases and Related Health Problems" 10th Revision
ICH-GCP International Council for Harmonisation of
   Technical Requirements for Pharmaceuticals for
   Human Use
IMP Investigational Medicinal Product

MHRA Medicines and Healthcare products Regulatory
   Agency

PI Product information
PRO Patient reported outcomes

QoL Quality-of-life

SAE Serious Adverse Events
SUSAR Suspected, Unsuspected Serious Adverse Events
TMF Trial master file

Addendum I

Handbook: The Duty for Sponsor Oversight in Clinical Trials – A practical guide.
Supplement to the e-book Edition 1 and 2 - Quality Report Template.

In overall, as outlined before, the conduct and reporting of the quality activities in a clinical trial and or development program should be provided in a consistent and transparent manner. A quality report template could be embedded into an overall guiding quality standard procedure. Below, a proposal regarding the content of a quality report will be shown.

Handbook: The Duty for Sponsor Oversight in Clinical Trials – A Practical Guide
Supplement to the e-book Edition 1 and 2 – Quality Report Template Risk-Based Clinical Data Review

Table of Content